HISTORY OF
CANALS

RICHARD TAMES

WAYLAND

Books in the series

History of Canals
History of Fairs and Markets
History of Food and Cooking
History of Toys and Games

Series editor: Sarah Doughty
Book editors: Dereen Taylor and Carron Brown
Designer: Michael Leaman
Production controller: Carol Stevens

First published in 1996 by Wayland Publishers Limited
61 Western Road, Hove, East Sussex BN3 1JD, England.

British Library Cataloguing in Publication Data
Tames, Richard, 1946 –
1. Canals – Great Britain – Juvenile literature 2. Canals – Great Britain – History – Juvenile literature
I. Title II. Canals
386.4'0941

ISBN 0-7502-1682-4

Picture Acknowledgements
The Boat Museum, Ellesmere Port 26; The Bridgeman Art Library, London/Fitzwilliam Museum, University of Cambridge 7, /Guildhall Library, London 34 (top), 38 (bottom)/Museum of London 9; Devon County Council Libraries 8; Dr Michael Essex Lowpresti 29 (bottom); Mary Evans 5, 11, 12, 13 (top); Greater Manchester County Record Office 14 (bottom); Hulton Deutsch Collection 14 (top), 15, 16, 22, 25, 29 (top), 31, 40, (bottom), 41; The Billie Love Historical Collection 20, 27, 30, 34 (bottom), 35, 38 (top); Archie Miles 36 (top); Museum of London 4; The National Waterways Museum, Gloucester 6, 13 (bottom), 18, 19, 21, 23, 28, 32, 33, 36 (bottom), 37, 39, 40 (top), 42, 43; Public Record Office of Northern Ireland 10; Wayland Picture Library 10 (bottom).

Illustrations on pages 5, 9, 17 and 24 by Peter Bull Art.

Special thanks to staff at the National Waterways Museum for their help and assistance.

Typeset by Michael Leaman Design Partnership
Printed and bound in England
by B.P.C. Paulton Books

Contents

Words that appear in **bold** in the text are explained in the glossary on page 46.

Water Transport

Transporting goods by water has always been popular. During the eighteenth and nineteenth centuries, canals helped industries grow. Today, they are used as popular holiday retreats.

Roman Times

Between AD 43 and 475, the cheapest and often the fastest way to move heavy goods was by water. The stone for the walls of Roman London, built around AD 200, was brought from Kent up the River Medway, and along the River Thames.

▲ *This museum model shows what London's riverside looked like in Roman times.*

The Middle Ages

Between 1066 and 1500, the great expense of moving heavy goods meant that most buildings were made of the materials available nearby. Areas without near supplies of stone used wood instead. In later centuries, houses and barns in these areas were made of brick. Usually, the only stone buildings were castles or churches. Heavy church bells and **millstones** were also very expensive to transport.

When William the Conqueror had the White Tower built in London, around 1070, water transport was used to bring strong stone all the way from Caen in Normandy.

Water Versus Road Transport

In the 1770s, the Scottish writer, Adam Smith, saw how the cargo of a single ship could be carried over 640 km from Leith to London in four days with a crew of just six or eight men. To carry the same load by road would need '50 broad-wheeled wagons, attended by 100 men and drawn by 400 horses'. The journey by road would take several weeks and be much more expensive.

▲ *A* ***packhorse*** *train. Notice how many* ***drovers*** *were needed to mind the animals.*

Nowhere in England is more than 110 km from the sea, and few places are more than 50 km from a navigable river. The coalfields of Northumberland, Durham and South Wales were all near the coast. Even the Midlands was not cut off because boats could sail up the River Severn into Shropshire and the River Trent into Staffordshire.

River transport was as cheap as coastal shipping but without the risk of losing cargo in a storm. The two main disadvantages to using rivers were that they did not always flow where they were needed most, and they were affected by drought and flood.

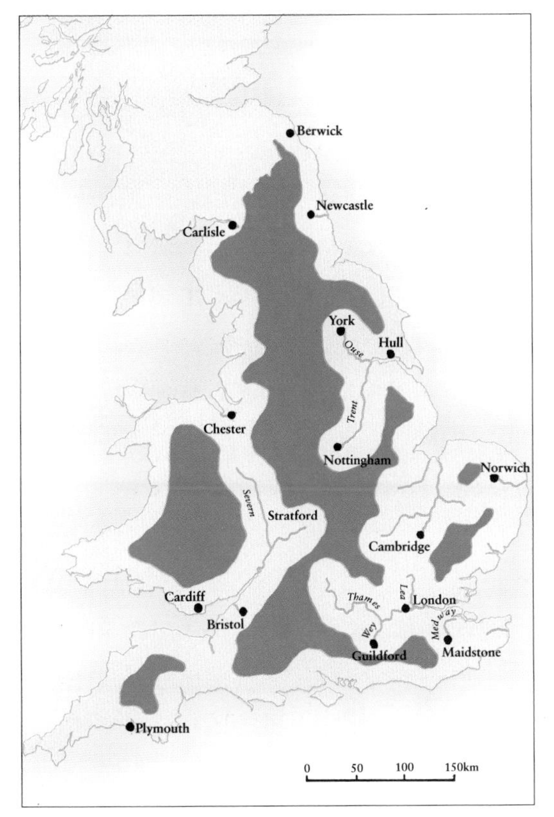

◀ *Navigable waterways, 1600. The shading shows those areas more than 24 km from navigable water.*

Using Water Transport

During the Middle Ages, water transport was also used for other heavy goods. Sometimes these were special objects that were very expensive to move, but were long-lasting and essential to a community, such as a church bell or millstone

One important industry which depended on water transport was the salt industry. Everyone needs salt as part of their daily diet, but salt is not found in every part of Britain. From the 1600s, Cheshire was one of the main centres of salt production, and salt was transported by sea and river to places as far away as London. Other goods that travelled long distances by water included iron and coal which were important to many industries, especially in the nineteenth century. Apart from being bulky, these products had one other thing in common – they would not rot or spoil on a long journey.

▼ *Some ports on navigable rivers were many miles from the sea – like Norwich, or Gloucester shown here.*

▲ *A riverside watermill in the fifteenth century. It was impossible for a barge to pass along a stretch of river where there was a* ***weir*** *or a watermill.*

Problems with Rivers

Rivers follow the landscape winding backwards and forwards rather than flowing in a straight line. They are often deeper along some stretches than others, which limits the size of boat that can travel along them. From time to time, rivers can become clogged with plants and rubbish, or overflow their banks. It was sometimes worth **dredging** riverbeds or digging out river banks to make them straighter or stronger if it made it quicker and cheaper to transport goods.

The Lea Navigation

During the 1500s, the population of London grew from 50,000 to 200,000. In order to feed such a huge number of people, corn merchants had to travel further and further away to buy enough wheat. This greatly increased transport costs. Most of the wheat from the area to the north of London was collected together at the market town of Enfield.

Between 1571 and 1581, the River Lea, a **tributary** of the River Thames, was made navigable as far as Ware, in an attempt to cut transport costs. Ware was much further from London than Enfield, but transporting grain by river barge actually worked out cheaper than bringing it by packhorse.

The First Canals

▲ *In 1830, a new basin for mooring boats was opened in Exeter. The flags show that a member of the Royal Family was attending this important occasion.*

The First Lock

Exeter, in Devon, is joined to the small seaport of Topsham by the River Exe. Between 1564 and 1566, local people dug a 6 km canal beside the River Exe to make a better link between Exeter and Topsham. Unfortunately, the canal was very shallow. Most ships continued to unload at Topsham, until the canal was deepened 100 years later. The Exeter-Topsham Canal used a special kind of **lock**, called a pound-lock, for the first time. Pound-locks enabled boats to change from one water-level to another.

A pound-lock worked by cutting off the flow of water with two sets of gates. The level of water in the section trapped between the two gates could be raised or lowered by opening and closing the gates one at a time. Before pound-locks were invented, the flash-lock was used. This dammed up a volume of water, and then released it in a single 'flash' to float a boat over shallow water. The flash-lock wasted water and involved long delays while enough water built up. The pound-lock wasted less time and water, and made it possible for an artificial waterway to cross sloping ground. It was also much easier for boats to operate on tidal waterways.

Improving Rivers

Britain had about 800 km of naturally navigable rivers. Between 1600 and 1730, the length of navigable river more than doubled to 1,856 km. This was achieved by dredging, straightening and removing obstacles, such as fish-traps and **weirs**. The money for these schemes was put forward by groups of businessmen, who got their investment back by charging **tolls** on goods carried along the improved stretch of river.

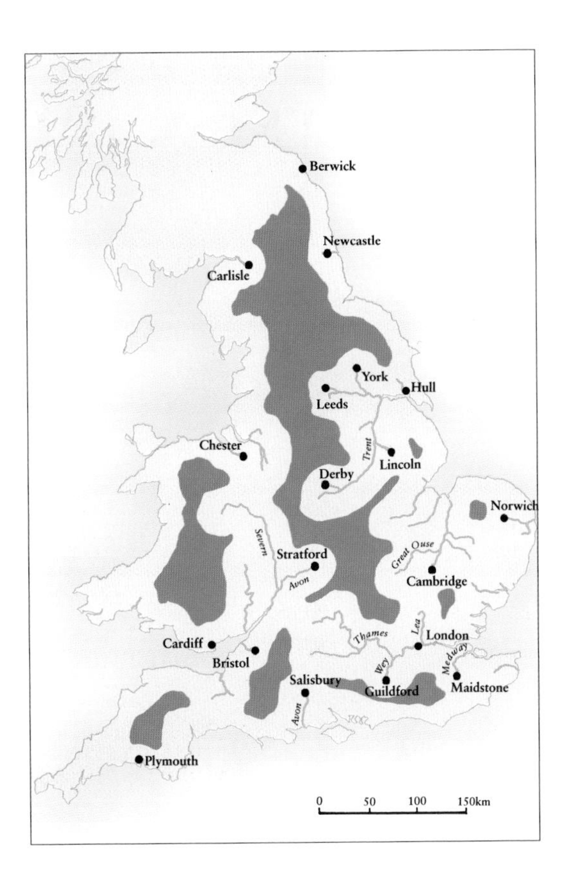

◀ *Navigable waterways, 1725. The shading shows those areas more than 24 km from* ***navigable*** *water.*

Britain's improved river system benefited all kinds of trades, especially grain and coal. A packhorse could not carry more than 120 kg, whereas the same horse could drag a load of 3,000 kg on a river-boat. On the River Thames and River Severn, barges with sails could carry loads of over 100 tonnes.

◀ *Sailing barges were used to carry goods on the Thames, as shown in this eighteenth-century painting. Lambeth Palace is shown on the far side of the Thames.*

An Early Engineer

The first true, long-distance canal in Britain was built in Ulster, Northern Ireland, between 1731 and 1742. It ran 29 km from inland coal mines to the sea at Newry and had 14 pound-locks. The canal's engineer was Thomas Steers (1672–1750), who had worked on building docks in Liverpool and London, and improved the River Mersey and River Irwall.

▲ *A horse-drawn barge along the Ulster Canal.*

The Sankey Navigation

Steers' pupil, Henry Berry (1720–1812), built the first major canal in England. The original plan was to improve a small stream, the Sankey Brook, so that boats could use it to reach the Mersey. The boats would be able to transport coal cheaply from the mines at St Helens to the salt-boiling works in Liverpool. As the salt-boiling process required over a tonne of coal to boil enough sea-water to produce two tonnes of salt, a cut in the cost of carrying coal would reduce the price of salt.

◀ *A staircase lock. This rise of five locks crosses the Pennines on the Leeds and Liverpool Canal.*

Berry realised that the Sankey Brook would never be made wide and deep enough to carry a profitable trade, and it would be better to build a new canal. It would still be possible to use the Sankey Brook to provide a reserve flow of water in very dry weather, and as an overflow in very rainy weather.

Work on the new canal began in 1755, and most of the canal was opened in 1757. Further extensions were opened in 1762 and 1772. What had begun as the Sankey Navigation Canal ended up being known as St Helens Canal. It ran for nearly 13 km and had ten locks, including a staircase lock.

◀ *St Helens, Lancashire, in the late nineteenth century. Coal production improved the economy, but not the level of pollution.*

St Helens

Thanks to its canal, St Helens grew into a successful industrial town. In 1779, local coal was used to smelt copper brought by water from Wales. An iron foundry began in 1798 and by the 1820s, glass-making and soap-making had become important local industries. The town's population rose from 4,000 in 1821 to 11,800 by 1845. Glass-making is still the main industry in St Helens today.

The Duke's Canal

The first really famous canal was built for Francis Egerton, the third Duke of Bridgewater (1736–1803). The Duke owned farms and coal mines at Worsley, near the growing industrial town of Manchester. Like many wealthy young men at that time, the Duke of Bridgewater travelled abroad and saw canals in France, and The Netherlands.

Manchester was surrounded by coalfields, but the cost of carrying coal by cart or packhorse doubled the price of the coal by the time it reached the town. The Duke could have transported his coal to Manchester along the Mersey and Irwell Canal, but its owners were greedy and charged very high prices. If the Duke built his own canal, he would be able to sell his coal and farm produce more cheaply, and also challenge the Mersey and Irwell Canal owners' control of the area.

In 1757, the Duke had his twenty-first birthday and was legally old enough to control his own fortune. He immediately began to look for an engineer to build his canal.

◀ The third Duke of Bridgewater (1736–1803). The Irwell Aqueduct is visible in the background.

Mr Brindley

The man chosen to build the Bridgewater Canal was James Brindley (1716–72). Brindley was the son of a farm labourer and had very little schooling. However, he had been **apprenticed** to a **millwright,** and learnt how to build and repair windmills, and watermills. Brindley worked closely with John Gilbert, the man who looked after the Duke's estates, and Gilbert was put in charge of the day to day running of the project.

▲ *James Brindley (1716–72) with a transit. This was a tool used for surveying heights and distances.*

A Problem to Solve

Before Brindley could build a canal from the Worsley mines to Manchester, he had to overcome a major problem – the River Irwell. Brindley's daring solution was to build an **aqueduct** 200 m long and 13 m high, which carried the canal right over the river. Brindley also built an underground tunnel that went right into the Duke's mines, so that coal could be loaded on to waiting barges. Over the years, 74 km of canals were tunnelled through the mines at Worsley.

▼ *A barge entering the Worsley Basin on the Bridgewater Canal.*

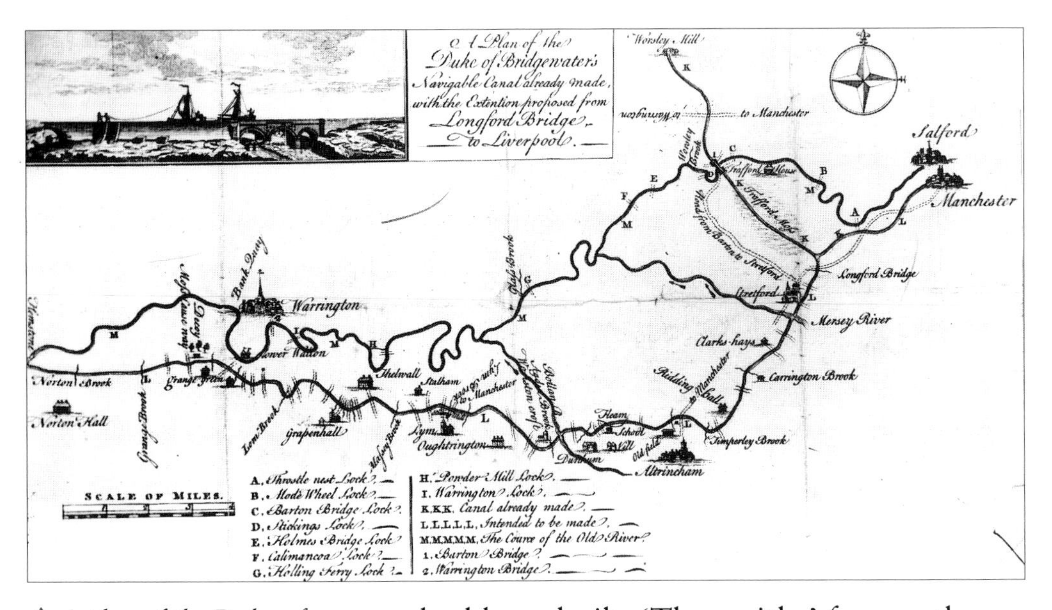

▲ *A plan of the Duke of Bridgewater's canal.*

Waste Not, Want Not

Thirty years after the Bridgewater Canal was finished, John Phillips wrote *A General History of Inland Navigation* (1793). In this book, Phillips described the clever methods by which the canal had been built: 'The smiths' forges, the carpenters' and masons' workshops...floated on the canal and followed the work from place to place. The Duke made the rubbish of one work help to build another. Thus the stones which were dug up to form the basin for the boats (near the coal-mine at Worsley) were cut into different shapes to build the bridges over the rivers, brooks or highways, or the arches of the aqueduct...'

Success!

When the first section of the Bridgewater Canal was completed in 1761, it was 16 km long. The Duke was able to cut the price of his coal by half and sold much more of it. In 1762, he persuaded Parliament to pass a law allowing him to extend the canal from Manchester to the great port of Liverpool, a distance of 48 km. This extension was completed in 1776.

▼ *Lord Ellesmere's pleasure boat being towed on the Bridgewater Canal around the year 1900.*

In 1767, a passenger service was started on the canal. The Duke of Bridgewater sailed his own pleasure boat, which was designed like the **gondolas** of Venice. He also allowed his friends to sail their boats on the canal.

The Wonder of the Nation

Visitors who saw the Bridgewater Canal were very impressed. In 1763, the 'Annual Register', a yearbook of important events, printed an eyewitness account of the canal's most famous feature: 'At Barton Bridge, he (Brindley) has erected a navigable canal in the air; for it is as high as the tops of the trees. Whilst I was surveying it with a mixture of wonder and delight, four barges passed me in the space of about three minutes, two of them being chained together, and dragged by two horses, who went on the terrace of the canal, whereon, I must own, I durst hardly venture to walk, as I almost trembled to behold the large river Irwell underneath me...'.

▼ *The wonder of the age. Barton Bridge Aqueduct allowed canal boats to pass over the River Irwell.*

The earliest canals had one limited aim – to link a coal-mine with a river, port or big town. But some far-sighted men looked forward to seeing the whole of Britain crisscrossed with canals, linking up rivers, ports and major inland cities. The first step was to join up the biggest rivers – the Severn, Thames, Trent, Great Ouse, Humber, Mersey, Forth and Clyde.

The Grand Trunk Canal

The Trent and Mersey Canal, also called the Grand Trunk Canal, was supported by various businessmen. These included the famous Staffordshire potter, Josiah Wedgwood, and the Birmingham manufacturer, Matthew Boulton. This 150 km canal would join the River Trent and the River Mersey, and pass through a number of important towns. It was hoped that other canals would be built to link in with it, like the branches of a tree sprouting from the trunk.

▲ *The great pottery-maker, Josiah Wedgwood (1730–95) was a keen supporter of canal building.*

Josiah Wedgwood saw that a canal would really benefit his pottery business by cutting the cost of bringing the best china clay from Cornwall. Even more importantly, smooth water transport would enable Wedgwood to ship out finished pots and plates from his factory with fewer breakages than if they were jolted in a cart, or on the back of a packhorse.

A Nearly National System

The completion of the Staffordshire and Worcestershire Canal, and the Birmingham Canal, made the Midlands generally (and Birmingham in particular), the centre of an impressive waterway system.

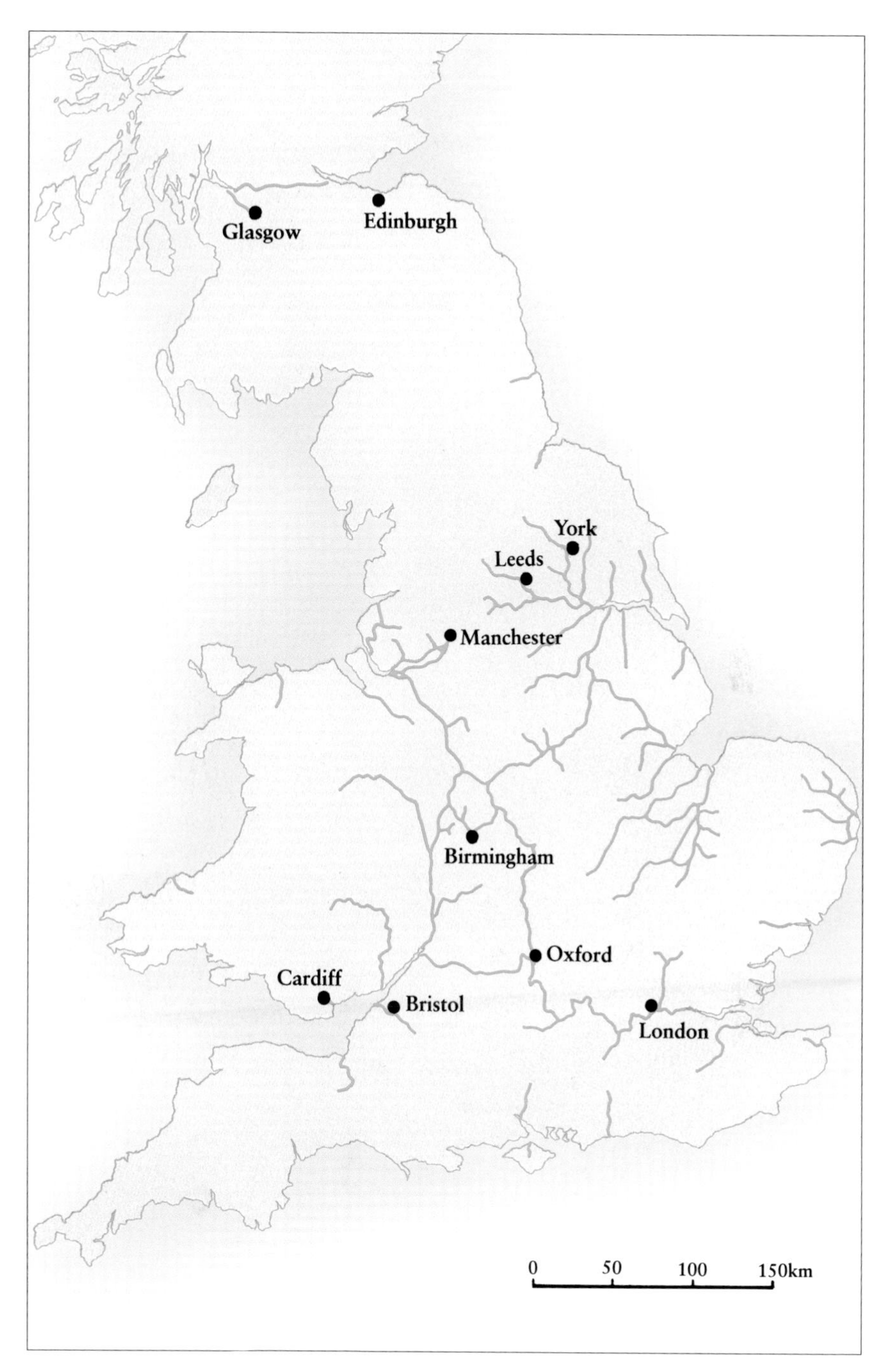

▲ *England's, Wales' and southern Scotland's inland waterways, 1790.*

By 1790, this system linked the major ports of Liverpool, Hull, Bristol and London with one another, and with important inland cities such as Oxford, Coventry and Stoke. In the north, work was started on canals to cross the Pennines from east to west.

The Leeds and Liverpool Canal, which was 330 km long, was begun in 1770, but not finished until 1816. The Huddersfield Canal, just over 30 km long, needed 74 locks and a tunnel 5 km in length. In Scotland, the most important project during this period was the linking of the River Forth and the River Clyde.

Canals were built across Britain at various times by different groups of people, and were often built to different measurements. Some boats were too wide to pass from one canal to another, which limited through-traffic. The system was, therefore, nationwide but not truly national.

Building canals cost huge sums of money, but could be very profitable. The Duke of Bridgewater's Canal cost £250,000 – at a time when a labourer on one of his farms would have earned just £20 a year.

Canal Companies

Most canals were built by local groups of businessmen and landowners. They formed a company which would persuade Parliament to pass an act to allow the canal to be built, raise money to pay for the work to be done, appoint an engineer and contractors to do the work, and then run the finished canal.

Few canal investors expected to make a quick fortune from their investment, although they did hope to get their money back. More importantly, investors expected their businesses, farms or mines to become more profitable as a result of the new waterway. Most early canals paid off because they were providing a much-needed route, or linking up two existing waterways. Shares in the most successful canals increased in value by between three and ten times. Rising values attracted speculators, who were less interested in long term profits from traffic than cashing in on the boom by buying and selling shares.

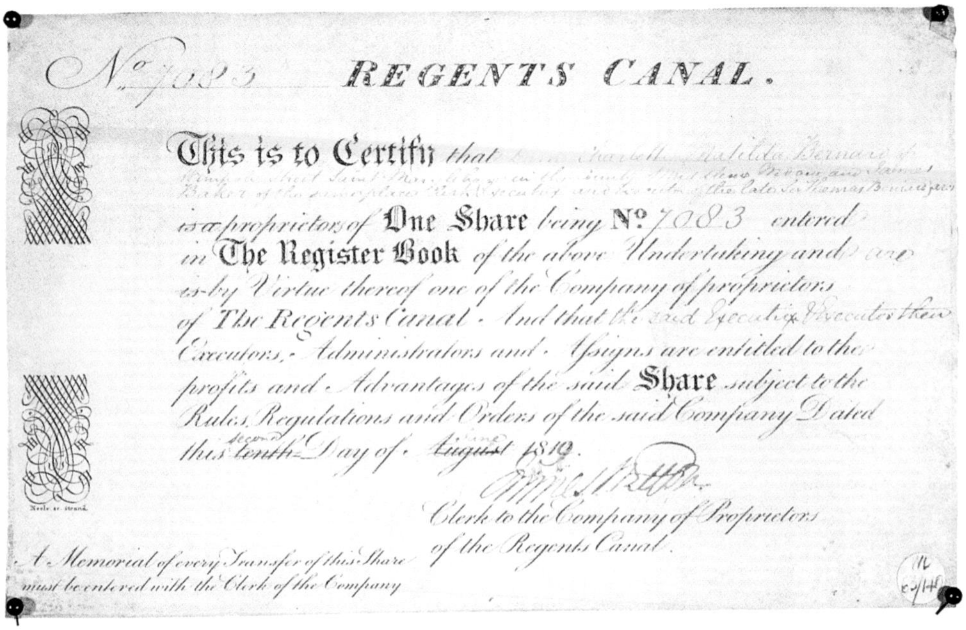

▶ *A share certificate held by an investor in the Regent's Canal.*

Canal Mania

The success of the early canals made many people think a canal built anywhere would be profitable. The country was gripped by 'canal mania' and between 1791 and 1796, Parliament approved 51 new canal projects. Some, like the Grand Junction Canal, between London and Braunston, and the Kennet and Avon Canal, linking London to the West Country, proved very worthwhile. But many did not. The long wars with France (1793–1815) left the country with high inflation and very short of money, which put an end to canal mania.

Winners and Losers

The building of a canal often cost more than was at first expected. The experienced James Brindley estimated the Leeds and Liverpool Canal could be built for £250,000. In fact, it cost £1,250,000, which was five times as much. When a company ran out of money before work was completed, it had to borrow more to finish the project. As a result, many canals had large debts to pay off before they even opened to traffic.

▲ *The Grand Junction Canal at Paddington in 1801, passing through open fields, and then past the outskirts of London.*

About 30 canal companies, mostly in the industrial Midlands, proved very profitable. However, almost 40 canals, including the few canals built in the agricultural south, where there were no mines or big cities, attracted so little traffic that they hardly made any profit at all.

Canal Builders

The men who built the canals were first known as 'navigators', which soon became shortened to 'navvies'. Building a canal needed two types of workmen – navvies and craftsmen.

Navvies

Unskilled labourers were needed to shift huge quantities of earth and rock. They had to be strong and used to working long hours in all weathers. A good navvy could shift 12 m^3 in a day – the equivalent to digging a trench a metre deep, a metre wide, and 12 m long.

In many areas, local farm workers and miners left farms and mines because they could earn more money digging a canal. When the work was finished, some went back to farm work or mining. Others became professional navvies and joined other canal projects. Large numbers of navvies came from Ireland and Scotland where good jobs were scarce.

▼ *A navvy ramming cobblestones into place to make a firm path.*

By the 1790s, at the peak of canal mania, there were at least 50,000 navvies at work. A project like the Lancaster Canal could employ 1,000 labourers, usually working in gangs of 50, and use 200 horses and carts, each with a driver. When the long wars against France ended in 1815, thousands of ex-soldiers looked for work navvying. Runaways from the law also joined navvy gangs, where no one asked 'Yorkshire Jack' or 'Thin Legs' who they really were.

◀ *Loading bricks for transport to the canal site. Wherever possible, bricks were made locally to save expense.*

Craftsmen

A canal was more than just a big ditch. Skilled craftsmen such as stonemasons, bricklayers, blacksmiths and carpenters were needed to build bridges, locks, warehouses and **wharves**. Most canals needed at least two bridges every 1.5 km. It was often necessary for canal builders to hire carpenters to make basic equipment such as cranes and wheelbarrows. Canal builders also provided their own kilns to make bricks, which were needed by the million.

Craftsmen were usually paid up to twice as much as navvies. They were also more likely to have their families with them, and to live in rented cottages. After the canal was finished, some craftsmen were kept on as maintenance men.

▲ *Employer John Pinkerton paid navvies working on the Basingstoke Canal with these tokens. Paying in tokens discouraged navvies from leaving the job because they could not be spent elsewhere.*

Tools and Equipment

Almost all work had to be done with simple tools – picks, shovels and wheelbarrows. There was also horsepowered winding gear to pull heavily loaded barrows out of ditches, too deep for ordinary wheelbarrows. Steam-engines were used for pumping water out of the bottom of the canal when it seeped up from the ground below, and blasting-powder was used to blow large rocks out of the way. When a canal had been dug out, its bottom and sides had to be 'puddled' – lined with clay up to a metre deep to stop the water draining out. This was usually done by cattle tramping up and down until the clay was thoroughly worked in.

▲ *A beam-engine, used to pump water from one level to another.*

Danger! Men at Work!

Accidents at work were common when men were moving heavy loads on muddy, slippery surfaces, or working with fingers and feet frozen with cold. There were no laws or government inspectors to protect workers from dangerous conditions, or force employers to provide a doctor or nurse to treat injuries. Tunnelling was the most dangerous task because of the danger of floods or roof collapse. Even if nothing went wrong, the labourers still had to breathe in foul air and gases from explosions.

'A labourer at the Lune Aqueduct had the misfortune yesterday to have three of his Fingers on the right hand taken off by the Piling Ram falling upon them. I should recommend him to the attention of the Committee to give him a small sum to assist him in his present situation.' (Archibald Millar, engineer to the Lancaster Canal Committee, 6 May 1794.)

A few companies did care about their workers – or the trouble unhappy workers could cause. The builders of the Caledonian Canal supplied their men with milk and oats at cost price. They even built a small brewery so that the men could drink beer instead of whisky.

Living conditions were also very primitive, with most navvies living in tents or huts made of turf, with no bathrooms or toilets. A navvy's life was hard, and navvies got a reputation for heavy drinking, fighting and rioting.

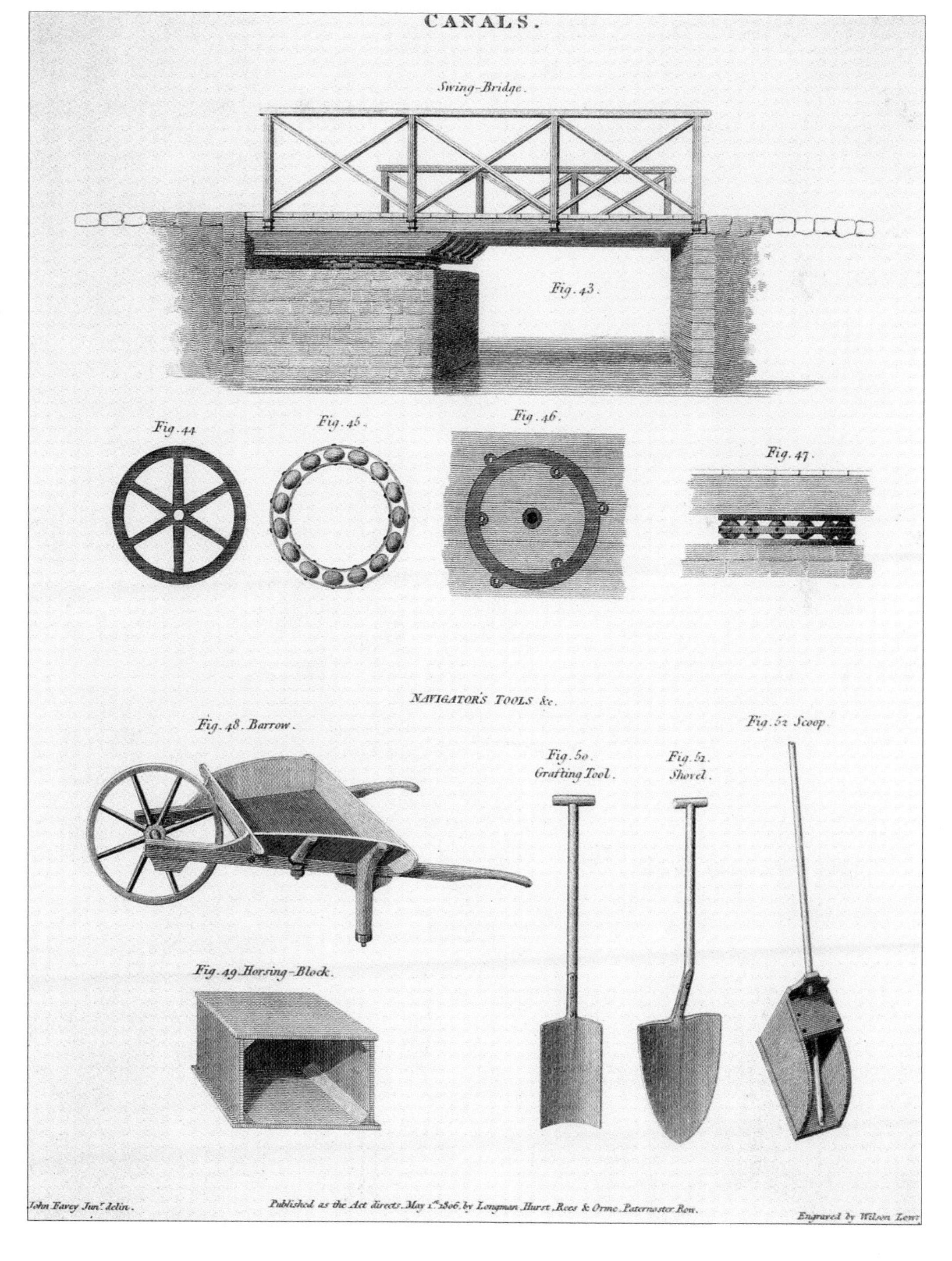

▼ *Most canals were built with simple hand tools like these.*

Amateurs and Professionals

The earliest canals were built by amateurs like James Brindley, who had to learn their business as they went along. In the last five years of his life, hard-working Brindley started no less than seven canals, totalling 540 km in length – but he never even learnt to spell properly!

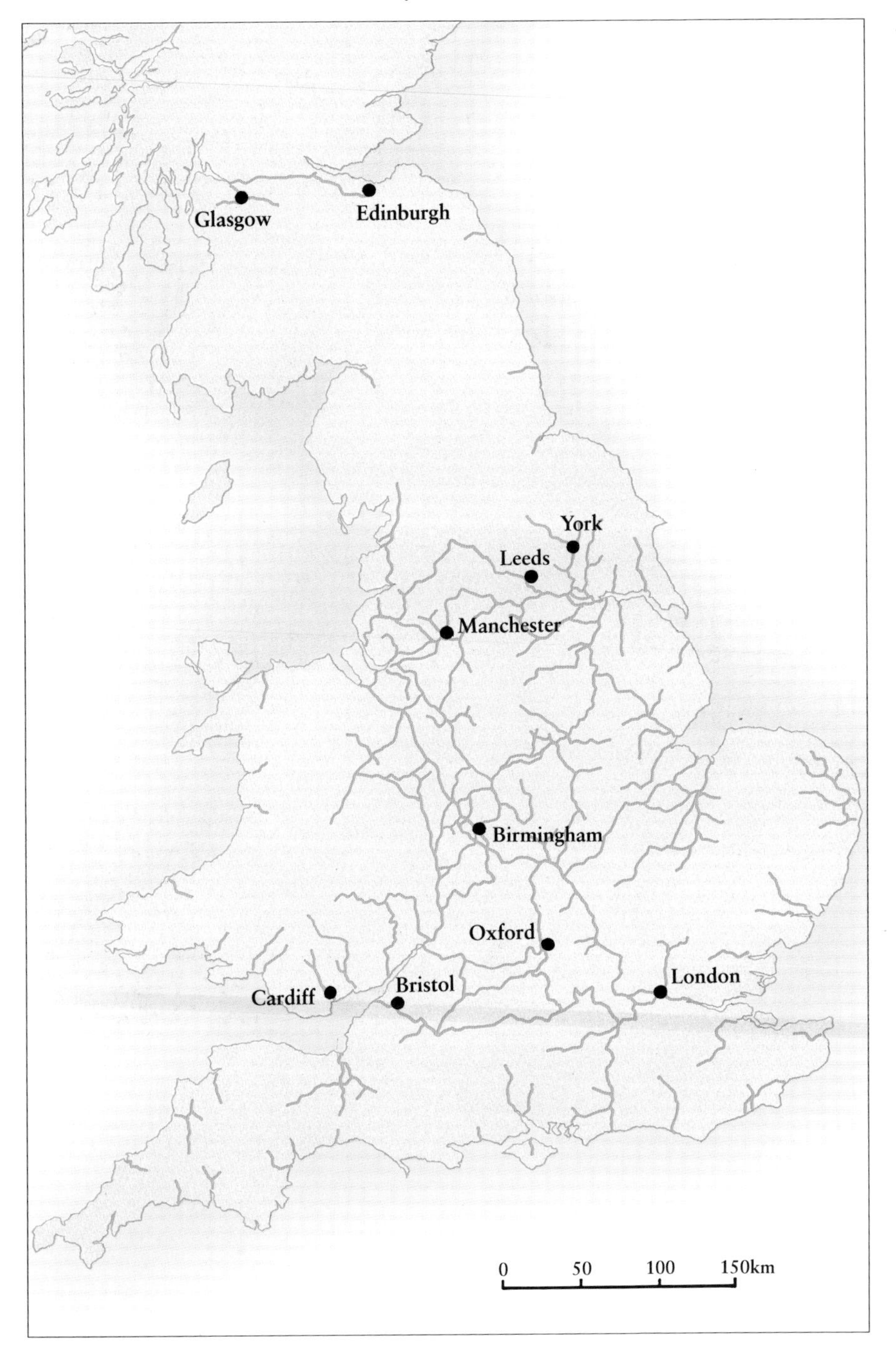

▲ *England's, Wales' and southern Scotland's inland waterways, 1850.*

Later, the building of canals involved difficult engineering problems, like building long tunnels or flights of locks. The experience gained by engineers on these projects was passed on to the next generation of canal-builders. In this way the new profession of civil engineering was created.

Masters and Apprentices

Many leading engineers learned directly from each other. Robert Whitworth who helped to join the Forth and the Clyde, was trained by Brindley. John Smeaton (1724–92), who designed the Forth and Clyde Canal, trained William Jessop (1745–1814), who designed the Barnsley, Erewash, Cromford, Nottingham and Ellesmere canals. John Rennie (1761–1821), who built the Crinan, Kennet and Avon, Rochdale and Lancaster canals, trained James Green (1781–1849), who built canals in south-west England.

Thomas Telford

▲ *Thomas Telford (1757–1834). The spectacular Pontcysyllte Aqueduct is visible in the background.*

The last great engineer of the canal age was Thomas Telford (1757–1834), who was the son of a poor Scottish shepherd. Telford first trained as a stonemason. He built 40 bridges in Shropshire between 1790 and 1796. Between 1795 and 1805, Telford and William Jessop built the spectacular Pontcysyllte Aqueduct to carry the Ellesmere Canal over the River Dee. The aqueduct was 38 m high and was supported by 18 cast iron arches mounted on stone piers. The water in the aqueduct was held in an iron trough. Telford pioneered the use of iron in the building of long aqueducts, because it could make much lighter troughs than stone. Iron troughs also needed less stones to support them. Telford also surveyed the route for the Caledonian Canal, and built the Shropshire Canal linking Birmingham to Liverpool.

Telford built 1,600 km of roads and 1,200 bridges, including the superb **suspension bridge** which joins the island of Anglesey to Wales across the Menai Straits. When the Institute of Civil Engineers was founded in 1818, to make sure all engineers had professional training, Telford became its first president. The new town of Telford, in Shropshire, is named in his honour.

The Industrial Revolution

The building of canals helped to bring about enormous changes in industry and how people lived.

Changes in Industry

Canals made it possible to use Britain's rich mines more efficiently. Between 1758 and 1802, 165 Acts of Parliament were passed to allow the building of canals. Of these canals, 90 served coal-mines, and another 47 served iron, lead and copper mines.

By cutting transport costs by at least half, canals lowered the price of building materials, coal and food. This helped industrial towns like Manchester and Birmingham grow into massive cities. In 1790, about 60,000 people lived in Birmingham. By 1841, this had risen to almost 150,000. The population of Manchester grew even faster – from 25,000 in 1772 to 181,000 by 1821.

▼ *Wedgwood's pottery works at Etruria were located on the edge of the Trent and Mersey Canal.*

Canal building increased the need for bricks, stone, timber and cast iron, creating more jobs in the industries that produced building materials.

Wages paid to navvies, and profits paid to canal shareholders were spent on beer, clothes, pots and pans and cutlery, boosting the industries that produced these goods.

Changes to the Countryside

The countryside also benefited from the canals. Cheaper transport costs meant that bricks, tiles and slates could be taken to areas that did not produce their own building materials. This helped to build better cottages. Many canals carried waste away from towns free of charge to keep them clean, and to use on fields as manure.

Huge changes occurred in South Wales as valleys rich in coal and iron were linked to the sea for the first time. The Welsh valleys became Britain's most important area for making iron and steel. By 1850, Merthyr Tydfil's Dowlais Works was the largest iron and steel producer in the world.

▲ ***Freight** traffic on the Regent's Canal.*

An Improved Transport System

By transporting bulky goods, canals helped to reduce the number of heavy carts using the roads. This protected road surfaces from being churned up. There were also fewer slow vehicles holding up fast stagecoaches. During the long wars against France, canals provided a safer way of transporting goods within Britain rather than sending them round the coast, where they could be attacked by French ships.

Inland cities benefited from the building of canals because businesses were able to send goods faster and more cheaply to other parts of Britain, and overseas.

Birmingham

When the agricultural expert, Arthur Young, visited Birmingham in 1790, he was amazed at the changes that had happened in the 12 years since his previous visit: 'The capital improvement since I was here before is the canal to Oxford, Coventry, Wolverhampton etc.; the port, as it may be called...crowded with coal barges...is a noble spectacle...this place may now probably be reckoned, with justice, the first manufacturing town in the world.'

A Frenchman, also writing in 1790, noted that: 'Until the middle of the century there was not one Birmingham trader who had direct contact with abroad. London merchants exported their goods. Now Russian or Spanish firms order what they want direct from Birmingham.'

By 1845, the 256 km of canals around Birmingham were carrying over 4 million tonnes of goods a year. Over half of this was coal, used to power the 700 steam-engines that were beside waterways.

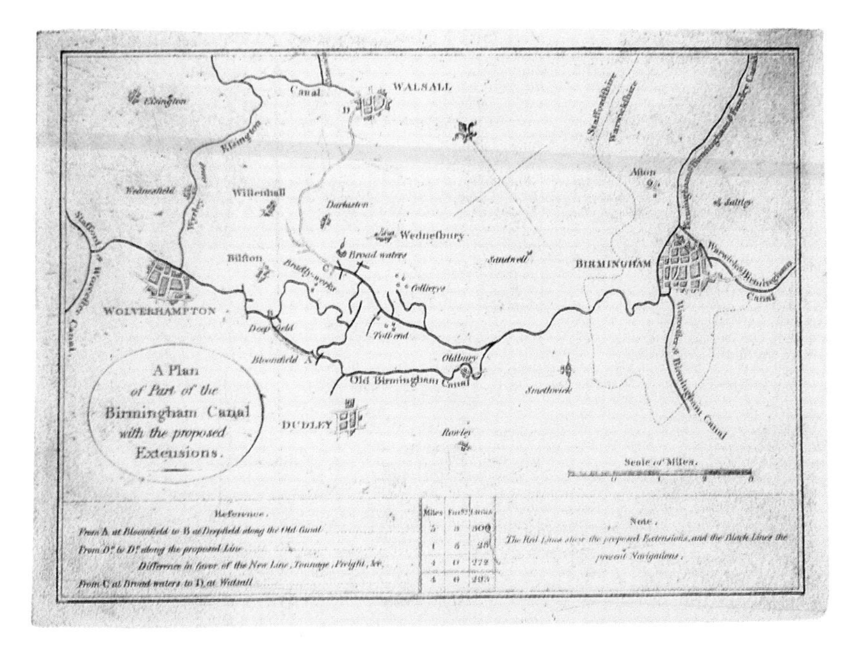

▶ *The Birmingham Canal system in the 1790s.*

▲ *Stourport in 1776. Notice how the whole town is centred around the canal.*

Kendal

The building of a canal could also have a huge impact on a small, sleepy town like Kendal in the Lake District. All sorts of businesses benefited with the opening of a link to the much larger town of Lancaster. The new wealth encouraged the people of Kendal to build a bridge, widen streets and put up rows of houses.

Stourport-on-Severn

Stourport, in Worcestershire, is Britain's best example of a town purpose-built to serve canal traffic. Apart from a large hotel and inns and warehouses for merchants and travellers, it also had works making iron, carpets, boats, leather and vinegar. Nowadays, Stourport trades on its canal history as a centre for people taking canal-boat holidays.

▼ *Stourport 200 years later – a popular stop for pleasure boats.*

Canal Towns

Canal traffic created jobs in towns where the canal linked other canals or rivers. Wharves, warehouses, boat-yards and rope-works were needed to keep boats and goods on the move. Such towns included Runcorn and Ellesmere Port in Cheshire, Shardlow in Derbyshire and Goole in Humberside.

Canal Life

Crews

Most long-distance boat crews consisted of two men on-board and a boy leading the towing horse. On canals where there was no **lock-keeper**, a third man sometimes opened and shut the lock gates.

▲ *Travelling by horsepower. The smoke would have come from a cooking stove, not an engine.*

Safety and Discipline

Fast packet boats, carrying parcels and passengers, had the right of way over boats carrying freight. Otherwise, overtaking was forbidden because fast-travelling boats damaged the canal banks and could cause accidents. If a boat sank, the boat owner had to pay for that part of the canal to be closed off and drained, so that the boat could be raised.

Compared with building canals, working on them was fairly safe. However, accidents did sometimes happen, especially in very cold or wet weather when it was difficult to hold on to things properly. Children were particularly at risk from being kicked by a horse, crushed in a lock or drowning. A 17-hour day was common, and many accidents happened simply because people were so tired. Another problem was smelly, dirty cargoes, such as bones, horn or manure, that carried a risk of disease, skin infections and bugs.

◀ *The keeper of Boulter's Lock, photographed in 1903. The lock-keeper's cottage is in the background.*

At first, boats were not allowed to go through locks or past a tollhouse at night. Lock gates were often padlocked, or a chain might be hung across the canal at night. Crews were forbidden to sleep aboard their boat or light a fire in their cabin. Drunkenness, stealing cargo and quarrelling with a lock-keeper or toll-taker could be punished by fines or dismissal from their jobs. Many canals were closed on Sundays and some companies made their boatmen go to church.

Tunnels

Where tunnels were too narrow to have a **towpath**, boats had to be 'legged' through, with the boatmen lying on boards laid across the deck and pushing against the walls of the tunnel with their feet. Very long tunnels often had gangs of professional 'leggers' who did this work full-time. It was usual for narrow tunnels to have one-way traffic rules, with set hours at which they could be entered by boats from each end.

▼ *It would have taken over two hours of hard work to 'leg' through this particular tunnel.*

▲ *A hard-working boatman's family pose for the camera.*

Canal Boat Workers

When canals were most important, around 1830, they employed perhaps as many as 100,000 people. Even in the 1870s, about 30,000 people worked on the canals. But the way of life for the canal workers had changed dramatically between the two periods.

Before the arrival of the railways, the canal-boat labour force was almost entirely male. Wives and children lived onshore, and many boatmen, particularly on the shorter canals, slept at home each night. But from the 1840s onwards, a new way of life developed. To cut costs, boatmen on narrow canals got rid of their crews and brought their families aboard to live with them. People who lived onshore called the boat families 'bargees' or 'water gypsies', but the families called themselves 'boaties'.

Inside a Cabin

As the narrow boat became a home, the cabin inside the boat was fitted with a coal-fired range for heating, cooking and boiling water. Smoke was carried off by a detachable chimney that could be taken down when the boat passed through low tunnels.

As the average cabin was only 3–4 metres long, everything inside was designed to save space. Coal for the range was stored under the step leading down from the deck to the cabin. The door of the food cupboard was hinged so that it could be used as a table. Children's beds doubled as seats in the daytime, when the parents' double bed was folded away. Brass oil lamps were used for lighting. Cupboards, water cans and crockery were often brightly painted with traditional designs showing roses, castles and boats.

▼ *A tin can for water, hand-painted in the traditional 'roses and castles' style.*

Canal Children

As boats were always on the move, the children of canal crews rarely had the chance to go to school. As late as the 1920s, probably three-quarters of canal children were unable to read or write properly. These children usually had little choice when they grew up but to continue with the same way of life as their parents. Boat people became a distinct sort of community, separate from people who lived settled lives ashore. They usually married other boat people who understood their way of life.

Canal Boats

▲ *Travelling on the Grand Junction Canal, 1801.*

It's Smoother by Water

Although canals were originally built to transport bulky goods, they also carried passengers. Travelling on a canal with large numbers of locks to pass through was obviously much slower than going by stagecoach. But where canals passed through fairly flat country they could manage roughly the same speed as a coach. Lightweight passenger boats, pulled by two horses changed every 6–10 km, could average 13 kph. Long journeys were usually very slow. For example, to travel from Wolverhampton to London took one week.

The cost of travelling by boat was less than travelling by coach. It was also less tiring for old and sick people than being jolted along a road. The smoother ride meant it was possible to read a newspaper or play cards. In bad weather, boats were much warmer and drier than coaches. Most passenger boats served food, but banned smoking and alcoholic drink. Many canals ran short daytrips to and from the nearest market town, or services which connected with a stagecoach line. Others ran special journeys for school outings, race-meetings, bathing or picnics.

▼ *Passengers travelling by stagecoach.*

Scotland

Passenger travel by canal was more popular in central Scotland than in the Midlands or South Wales. At first, passengers and goods travelled in the same boats. But from 1809 onwards, a daily 'passengers only' service began, with cabins equipped with newspapers, books, games and refreshments.

Passengers were carried on the Forth and Clyde Canal from its opening in 1790. In 1820, a steamboat service was started from Inverness to Fort Augustus on the Caledonian Canal. By 1822, steamers were using the Crinan and Caledonian canals to link Glasgow, Oban, Fort William and Inverness. By 1828, horse-drawn boats on the River Forth and River Clyde were covering the 39 km from Glasgow to Falkirk in three hours.

▲ *Passengers still travelled by steamer on the Caledonian Canal until 1939.*

In the 1830s, the Paisley Canal used specially built passenger boats, which travelled at 16 kph. Between 1812 and 1836, the number of passengers carried on the Forth and Clyde Canal rose from 44,000 a year to nearly 200,000. From 1831, it was possible to sleep through an overnight journey from Glasgow to Edinburgh, covering the 90 km in 11 hours. Steamer services continued on the Crinan Canal until 1929, and on the Caledonian Canal until 1939.

▲ *A modern narrow boat, adapted for cruising in comfort.*

Narrow Boats

Canal boats came in many shapes and sizes, but the most common was the narrow boat developed by James Brindley, to fit snugly inside the locks he designed for the Trent and Mersey Canal. The locks were 22.5 m long by 2 m wide, while the boats were 21 or 22 m long by 2 m wide. Narrow boats could carry an average of 20–30 tonnes, depending on the depth of water in the canal.

▼ *A steam engine hauls this coal-filled tub boat up a sloping railway.*

Tub Boats

Tub boats were about 6 m long and 2 m wide, and were much smaller than narrow boats. They carried loads of only 5 or 6 tonnes and were used to cross high ground instead of using flights of locks. A steam-engine hauled the tub boat up a sloping railway.

Barges

Broad canals, such as the Leeds and Liverpool, Rochdale, Grand Junction and Kennet and Avon, could carry wide boats or barges that were twice as wide as narrow boats. Barges could carry loads of 50–80 tonnes. They often had sails and hinged masts so that they could pass under bridges on larger rivers.

Iron and Wood

John Wilkinson, the iron manufacturer, thought iron was a miracle material. He had iron furniture in his house and even wore an iron hat! In 1787, Wilkinson built an iron barge and launched it on the River Severn to show that an iron boat could float. But until the 1830s, iron boats cost much more to build than wooden ones, and never really caught on – except for a few special ones built as icebreakers for use in very cold winters.

▼ *A broad-beamed barge used for carrying bulk freight (in this case stone), on the Leeds and Liverpool Canal.*

Steam Power

A few steam-powered boats were built, but were only popular on broad waterways where there was enough passenger traffic to make them profitable. The problem with steamboats was that their engine and fuel took up room that could have been used for cargo. Steamboats were also very expensive to build and repair if they broke down, and they needed a skilled operator to keep them working. Another problem with steamboats was tha they churned up the water and damaged the canal banks, which made expensive repairs necessary.

Victorian Times

▲ *George Stephenson's 'Rocket' could travel at almost 65 kph. A picture of the 'Rocket' appears on the back of an English £5 note.*

A Revolutionary Steam-engine

Many canals depended on the railways to link them with mines and quarries. Railway wagons were drawn by horses, or powered by stationary steam-engines. In the early 1800s, mining engineers such as Richard Trevithick and George Stephenson experimented with steam-engines called locomotives. These were powerful enough to move themselves and drag a load. By the 1830s, there were locomotives that could haul far more than the largest barge could carry – and ten times as fast.

Railways Link the Cities

The first railway with locomotives to compete directly with a canal was the Liverpool and Manchester Railway, opened in 1830. Railways, like the early canals, proved hugely profitable, and there was 'railway mania' during the 1830s and 1840s. By 1850, all Britain's major cities were joined by rail and there were over 9,600 km of track, making the railway network already almost one-and-a-half times as long as the waterways.

▶ *The London to Greenwich railway line crossing the Surrey Canal around 1840.*

Railways Versus Canals

In 1840, the Kennet and Avon Canal had an income of £48,269. This income was because the canals had been carrying materials for the building of the Great Western Railway, which ran parallel to the Kennet and Avon. By 1843, when the railway was fully open, the canal's income had fallen to £32,045.

Canals tried to meet the competition from railways by drastically cutting their freight rates, sometimes to as low as one-seventh of what they had been. They also tried to offer a better service by operating on Sundays and through the night.

Some railway companies responded by buying up small country canals. At the height of 'railway mania', between 1845 and 1847, 28 waterways, representing one-fifth of the length of the entire network, were taken over by railway companies. Sometimes the railway companies just let the canals fall into ruin, not dredging them out regularly, repairing their banks or bothering to advertise for business.

◀ *Gloucester Docks, around 1900. Goods could be unloaded from barges directly into warehouses on the canal side.*

A number of canals in the industrial Midlands and North had one major advantage over the railways – it was possible to unload goods direct to factories on the canal side which meant that there were no extra transport costs.

▶ *The proposed route of the Manchester Ship Canal from Liverpool Docks to the centre of Manchester.*

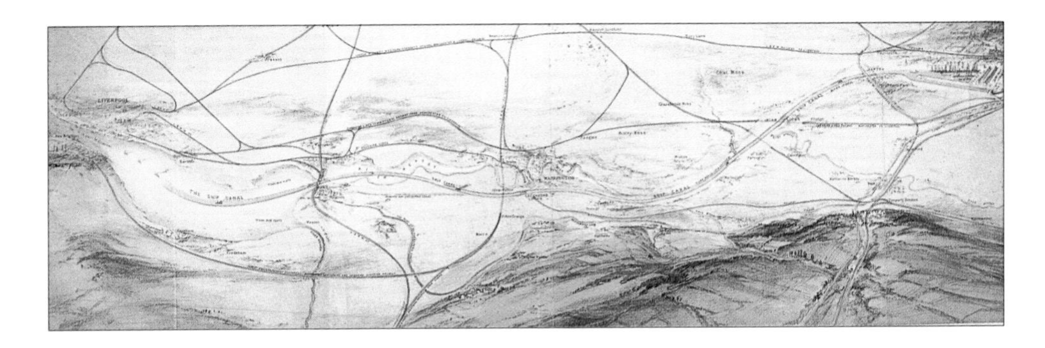

The Manchester Ship Canal

For 50 years, the mills of Manchester, Britain's largest cotton-manufacturing centre, relied on the Liverpool and Manchester Railway to bring raw cotton from Liverpool docks. But many manufacturers dreamed of 'bringing the sea to Manchester'.

In 1882, a group of Manchester businessmen asked Parliament for permission to build a canal big enough to carry ocean-going ships of 12,500 tonnes into Manchester. For three years, Liverpool businessmen, who feared that ships would no longer use their port because of the canal, argued against the plan. But Parliament finally gave permission in 1885. Work on the biggest British civil engineering project of the nineteenth century began in 1887.

▼ *The building of the Manchester Ship Canal. Notice the number of steam-powered machines in operation.*

Building the Manchester Ship Canal took seven years, required 70 million bricks, 750,000 tonnes of granite and 46 km of huge timbers. It employed 16,361 men and boys, assisted by 6,300 wagons and the largest concentration of steam-powered machinery in British history. This included dredgers, pumps, cranes and 173 locomotives running on 365 km of temporary railway track – all to remove more than 50 million m^3 of soil.

A New Aqueduct

Construction of the Manchester Ship Canal involved changing the course of the River Irwell and the River Mersey, diverting five railways, building seven swing-bridges and demolishing Brindley's historic Barton Bridge Aqueduct to replace it with a modern one. For all these reasons, plus a disastrous flood in 1890, the project finally cost more than twice the original estimate of £7 million. After the flood, the city of Manchester rescued the canal company with a loan of £5 million, and took control of the project.

▼ *At the opening of the Manchester Ship Canal, in 1894, an elegant yacht passes a new swing-bridge that replaced Brindley's famous Barton Bridge Aqueduct.*

Success for Manchester

Manchester had made an excellent investment. In 1894, the first year of operation, the Manchester Ship Canal carried 686,158 tonnes of ocean traffic and 239,510 tonnes of barge traffic, paying £97,901 in tolls. Twenty years later, barge traffic was still only 322,943 tonnes, but ocean traffic had risen to 5,457,218 tonnes and toll income to £654,937. Manchester had become Britain's third biggest port, and Trafford Park, on the edge of the canal, was Europe's first and largest industrial estate.

Canals Today

▲ *A steam-dredger helping to clean up the Basingstoke Canal.*

Conservation and Recreation

Although over 2,000 km of Britain's canal network is still navigable, its importance to industry has long passed as railways became more important. Nowadays, the canals' main value is for recreation and conservation. Cruising in a narrow boat can be a very pleasant way to see the countryside, although passing through locks makes sure that even lazy holiday-makers have to get some exercise. The Shropshire Union Canal is very popular for holiday cruises. Even ruined parts of canal, choked by weeds, serve a purpose as refuges for birds and other kinds of wildlife.

▶ *A family enjoying a narrow boat holiday.*

Canal Museums

▲ *The National Waterways Museum at Gloucester Docks.*

The heritage of the canal age can be seen at more than a dozen different museums. Many are housed in buildings that were once part of a working canal. The canal museum at Devizes, in Wiltshire, is set in an old granary, dating from 1810. At Linlithgow, in Scotland, the canal museum is in a former stable for canal horses. At Llangollen, in Wales, the museum is in a warehouse built by Telford, and at Reading, in Berkshire, it is in a pumping station. At Pontymoel, in Wales, where the Monmouthshire and Brecon and Abergavenny canals meet, there is an old toll-keeper's cottage, built in 1814.

Some canal museums have special features. The National Waterways Museum at Gloucester has a replica of a canal maintenance yard, a stable with horses and a working steam-dredger. At the Waterways Museum at Stoke Bruerne, Northamptonshire, there is a full-size replica of a canal boat cabin, with its cabin decorated traditionally. At Bath, the Canal Study Centre is housed in 'Bodmin', a narrow boat built in 1936, which was carrying coal on the Grand Union Canal until 1976.

One of the largest canal museums is at Ellesmere Port, in Cheshire, where the Shropshire Union Canal joins the Manchester Ship Canal. Here, cargoes were once transferred from ocean-going ships to the smaller boats used on canals. Now the museum occupies the old warehouses and workshops. There are over 50 boats in the museum's collection, including wide and narrow boats, steam-powered tugs and even icebreakers.

Timeline

1564	1571-1581	1742	1757	1761
Exeter-Topsham navigation includes the first pound-lock.	Improvements to the River Lea.	Newry Canal opened.	Sankey Brook Navigation (St Helens Canal) opened.	Duke of Bridgewater's Canal from Worsley to Manchester opened.

1780	1790	1791–1796	1803
Passenger traffic begins in Ireland along the Grand Canal.	River Mersey joined to River Trent and River Thames.	'Canal mania' reaches its height as Parliament authorizes the building of 51 new canals.	Work begins on the Caledonian Canal.

1923	1925	1930s	1946	1948
A successful strike confirms canal-boat workers' rights to join trade unions.	Canal traffic stopped for several weeks due to severe winter weather.	Diesel engines replace steam-engines and horses for power.	National minimum wage agreed for canal-boat workers.	Canals taken under government control.

1767	1772	1776		1777
Passenger traffic begins on the Bridgewater Canal.	River Mersey linked to River Severn at Stourport.	Bridgewater Canal connected to River Mersey.		River Trent connected to River Mersey.

1845	1858	1877	1894	1906
'Canal Carrier Act' enables canal companies to compete with railways by carrying parcels and freight.	Inland waterways reach a peak of 6,800 km.	The Canal Act regulates hours of working and living conditions on canal boats.	Manchester Ship Canal opened.	First narrow boat fitted with a diesel engine.

1958	1964
Last narrow boat to be built entirely of wood.	Restored Stratford-upon-Avon Canal reopened.

Glossary

Apprenticed
Employed for a set time, usually on a low wage, to learn a trade.

Aqueduct
An artificially constructed channel, usually in the form of a bridge, which carries water over long distances.

Cargo
Goods carried by ships.

Dredging
Clearing a waterway by removing mud and other obstacles from the riverbed.

Drovers
People who drive herds of cattle to markets.

Freight
Goods to be transported from one place to another.

Gondolas
Boats used on the canals of Venice. They were powered and steered by a man with an oar standing at the back.

Lock
A section of a canal that can be closed off by gates to raise or lower the water-level so that boats can pass through.

Lock-keeper
A person who is responsible for working the locks on a section of canal.

Millstones
Heavy, flat stones used by millers to grind grain.

Millwright
A worker skilled in making and repairing windmills and watermills.

Navigable
Suitable for boats to go on.

Packhorse
A horse used to transport goods.

Suspension Bridge
A bridge that has a road or walkway attached to cables. The cables usually hang between two towers at either end of the bridge.

Tolls
Charges paid to a canal owner when goods are moved on the canal.

Towpath
A path running alongside a canals where horses can be led to drag barges along.

Tributary
A river or stream flowing into a larger river or lake.

Weir
A dam built across a river to raise the level of water upstream or to regulate its flow.

Wharves
Platforms on the edge of rivers where ships may load or unload goods.

Books to Read

Children

Oxlade, C: *Canals and Waterways* (Watts , 1994)

Siliprandi, K: *Victorian Transport* (Wayland, 1993)

Wilson, Dr A: *Transport* (Dorling Kindersley, 1995)

Adults

Hadfield, C: *Illustrated History, British Canals* (David & Charles, 1984)

Paget-Tomlinson, E: *The Illustrated History of Canals and River Navigations* (Sheffield Academic Press, 1993)

Smith P L: *Discovering Canals in Britain* (Shire Publications, 1989)

Index